INTRODUCTION

The 2020's will go down in history for the rise of the robotic workforce. Service robots will be a key helper in many homes. For example, Samsung is developing Ballie, a household/companion robot that is the size of a tennis ball that's loaded with artificial intelligence. It's designed to understand, support and react to the owner's needs. Incredibly, it can connect and control smart devices in the home to the owner's specifications. As for pets, there will be plenty of little robots like Adorable Kiki and Aibo to choose from. Some of them will sit and stay on command. These are some of the many, fascinating household robots that are showcased in my book "New Robots for the 2020's.

Health Care Robots

One of the biggest megatrends in robotics is the use of health care robots which is greatly accelerating because of the COVID-19 pandemic. An example, included in the book, comes from Wuhan, China where a smart field hospital is totally run by health care robots, controlled by medical staff in a safe and remote location. In Singapore, robots are taking temperatures, dispensing hand sanitizers and encouraging social distancing. UV Light Robots are in high demand to deep-cleanse public places of the highly contagious COVID-19 virus. Venues include the hospitality industry, public transportation and much more. With COVID-19 likely to be a global health concern for some time, the demand for health care robots is expected to dramatically rise during this decade.

Autonomous Mobile Robots

The market for autonomous mobile robots is expected to hit $221 billion by 2030, growing from $29 billion in 2019. We showcase the latest including Misty 11 Robot Concierge, Spot Robodog Internet Sensation, Robots that deliver door to door, China's restaurant robots and Italy's very strong robot that can haul a plane. We also delve into the booming global drone business which is expected to be a $120 billion market by 2030.

5G Connected Robots

This decade robotic technology is being revolutionized with the inclusion of 5G systems that make robots more responsive, connected, fast and highly advanced. The 5G connectivity for robots is empowering and will provide any robot, from a $200 personal robot to a $2 million industrial robot with the key capabilities they need: sensing, thinking, acting and communicating. This book is an exciting look at the future of robotics in our personal/ business lives and in global industries during this decade.

TABLE OF CONTENTS

AUTHOR'S BIOGRAPHY

Ed Kane created and serves as Executive Producer of CEO Global Foresight, a national news program on PBS focused on breakthrough innovations. He is the author of 20 books on the latest innovations across industries. Ed is a science graduate of the University of Pennsylvania.

1. 5G Connectivity for Robots & Drones

Source: Qualcomm's 5G Robotic Connectivity System

New Tech from Qualcomm

San Diego-based mobile chipmaker Qualcomm has introduced revolutionary new technology for robots and drones. It is the first 5G system with hardware, software and tools for advanced robots. It's called the RB5 system. Empowering robots and drones with 5G connectivity takes them to the next level of connectivity, quickness and responsiveness. 5G is not just for smartphones anymore!

New Enabling Technologies

RB5 uses new Qualcomm technologies. For instance, it uses the company's new QRB5165 processor for robots. And it has a companion module for 4G LTE and 5G connectivity through Qualcomm's X55 modem, which is also used in smartphones.

Creativity Built-In

The RB5 has a tremendous amount of artificial intelligence built-in to enable robot and drone developers to create machines that are more responsive, smart and fast. Qualcomm expects this technology to reach into the consumer, business, industrial, defense and professional services markets for more customized, smarter, faster robotic solutions that are easy to use.

One-Stop Robotic Shopping

According to Dev Singh, who is the head of Qualcomm's Robotics, Drone and Intelligent Machine Business, any robot - whether it's a $200 personal robot or a $2 Million industrial robot - needs these critical capabilities: sensing, thinking, acting and communicating. Singh says Qualcomm's processor RB5 "packs all of that together".

This is a look at the future of robotics in your personal and business life.

2.Robot Lawn Mowers: Tech At the Cutting Edge

Source: Worx

Robot, Mow My Lawn!

Don't want to mow your lawn during summer heat waves? Let your robot lawn mower do the job for you! A growing number of people are doing just that. According to Worx, the company that manufactures the highly rated Landroid robot mower, autonomous robot lawn mowers are growing in popularity.

Loaded with Technology

There are a growing number of robot lawn mowers on the market. Among the top rated are Flymo, Worx's Landroid, Lawnmaster, Robomow, Honda's Miimo and Husqvarna Automower. The robots use artificial intelligence and sensors to avoid hitting obstacles. A special perimeter wire installed around the lawn keeps the mower on the owner's property. The mowers are battery operated, quiet and run automatically, based on the schedule that the owner customizes in an app. And the companies say they are safe with children and pets.

Buy or Rent

Still, the robot lawn mowers are pricey. They average $1,000, with some models as high as $3,500 and others as low as $800. There is another option. You can rent the robot mowers. For a monthly fee, some companies will set up an automatic mowing system at your home and do maintenance on the machine if any problems arise.

Green Technology

Letting the robot do the lawn is a great use of green technology at the cutting edge. The lawn is cut to the owner's specifications and the mower dutifully returns to its docking station to get recharged for its next mowing session. It's technology helping to make for the perfect summer.

3.Baseball's Robot Umpires

Source: South Korea's Robot Umpire

South Korea's Automated Ball-Strike System

South Korea's minor league baseball teams start utilizing robot umpires during regular games in August 2020. About 20 games in the Korean Baseball Organization's Future League have the automated ball-strike systems operating for the entire 2021 season. If it works as expected, South Korea's Major League baseball teams will deploy it, too, for their 2022 season.

Advanced Radar Technology

The Korean Baseball Organization (KBO) calls the system the "robot umpire". But, there are actually no robots behind the catcher on the field to make the calls. This is a radar-based ball tracking system that is automated, pinpoints the location of the pitch and tells the human home plate umpire, via an earpiece, the correct call. The ball's trajectory is shown on a large screen display at the ballpark.

Major League Baseball's Electronic Strike Zone

In the United States, Major League Baseball experimented with the electronic strike zone at the minor league level with eight teams in 2019. They liked the system and wanted to expand the automated system to other minor league teams. But the pandemic hit and put baseball on hold in the spring and into the summer of 2020. Nonetheless, the games will resume. They are in South Korea. And, South Korea's grand experiment with robot umpires is another great example of automation and robotics becoming part of our favorite pastimes. And in this case, taking the element of "human variables" out of the volatile equation of sports.

4. Spot RoboDog Internet Sensation

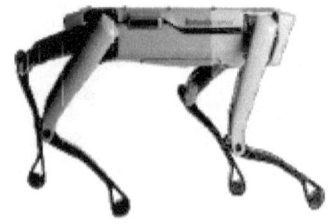

Source: Boston Dynamics

On the Market & Priced at $74,500

Boston Dynamics claims that their robot dog Spot is the most advanced mobile robot in the world. Their internet-famous Spot is now available for sale. Boston Dynamics has been developing Spot for years. Until now, only "select" clients could lease it. The price-tag to buy Spot is $74,500. The robot is designed for inspections, sensing and remote operations. Its big market opportunities are public safety, healthcare, construction, commercial and industrial uses.

Top Pedigree

Spot is a highly advanced and sophisticated robot. It's been used in Singapore to encourage social distancing during the COVID-19 pandemic. During a trial run, it created 3D maps of construction sites. It can dance, open and hold doors. Boston Dynamics says it navigates autonomously, handles rough terrain and avoids obstacles.

Top Dog

Thus far, the primary customers purchasing Spot are commercial and industrial companies which, for instance, have sensors that need to be placed in dangerous locations where they don't want to put people in harms' way. And, companies are using Spot for highly repetitive tasks. Interestingly, Boston Dynamics is working on advancing its teleoperations in order to enable customers to give remote demonstrations of Spot on duty. Also, Spot is small enough to be used indoors.

5.Samsung's Ballie Robot

Source: Samsung

Small Robot with Big Tech

Samsung has invented a life-companion robot that is the size of a tennis ball. The name is Ballie. The robot is designed to understand, support and react to the needs of the owner. Ballie is small, round, rolls and is customized to work in households. Samsung says it can function as a security robot, senior helper, fitness assistant and friend to kids and pets. The robot responds to commands, like "Come" with chime sounds and obeys.

Connectivity With Other Smart Devices

This little robot has big technology powering it. It has built in artificial intelligence capabilities. Ballie uses sensors and data that it gathers within the home to create personalized, immersive experiences for the owner. Incredibly, the tiny robot can connect with and control other smart devices in the home. This is cutting edge robotic technology designed to help improve the daily life of the owner. And it's the size of a tennis ball! Samsung is continuing to build Ballie's capabilities to include functions like turning on the TV, opening the curtains and commanding a robot vacuum to clean the rugs. Samsung has not yet set a release date or price.

6.UV Light Robots Fighting COVID-19

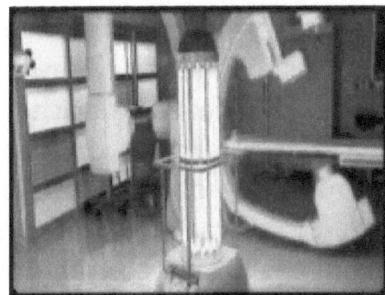

Source: UV Light Robot in Hospital

Hotels, Public Transit, Amazon Putting UV Robots to Work

Ultraviolet light (UV)) has been used as a disinfectant in hospitals for decades. As businesses start re-opening after the COVID lockdowns, the use of UV is rapidly expanding into the hospitality industry, public transportation and beyond including entertainment venues.

Beverly Hills Hilton

An example is the Beverly Hills Hilton Hotel, where 3 UV robots from San Antonio-based Xenex Disinfection Services are on deep cleansing patrol. Xenex has UV robots deployed in more than 500 hospitals around the world, including the Mayo Clinic. In two minutes the high energy pulses of UV light from the robots eliminate dangerous viruses like COVID by 99.999%, according to the company and independent analysis by the Texas BioMedical Research Institute. UV works at the speed of light and provides a rapid, thorough cleansing that is chemical free. The COVID virus can survive on surfaces for several days, spreading it to anyone who touches it. The UV light robots work in secured spaces without humans present. Direct human exposure to UV light can harm the eyes and skin.

Kennedy and a Global UV Robot Workforce
Xenex's UV robot Kennedy at the Beverly Hills Hilton cleans surfaces throughout the hotel, disinfects guest rooms and cleans guests' luggage and packages. The purpose is to keep the staff and guests safe from the COVID-19 virus. The Intercontinental Hotel Chain is also using UV robots. New York City's subway system has a pilot program using UV portable lamps to clean the subway cars. Amazon has developed a UV robot and has tested it at their Whole Foods Chain. And, PBA Group, a robotics company in Asia, has dispatched teams of UV robots "Sunburst UV robots" to clean malls across Singapore. A rising robotic workforce of UV robots is being deployed to protect humans from the highly contagious COVID-19 virus. It's an important robotic megatrend.

7.Telepresence Robots and Classrooms

Source: Oregon State U Engineering Professor & Telepresence Robot

Source: Graduation by Telepresence Robots

New National Science Foundation Research
A new study by the National Science Foundation has found that telepresence robots help students, who are learning remotely, feel more a part of the classroom. COVID-19 has forced millions of students out of the classroom and into remote, distance learning classes. The National Science Foundation finding is an important one and underscores another important role for telepresence robots to play: assisting and optimizing distance learning.

Learning Formats
The National Science Foundation, joined by researchers at Oregon State University, examined student experiences in three different circumstances: in-class learning, learning remotely through a telepresence robot and distance learning by calling into class, livestreaming or recorded class sessions.

Remote, Robotic Learning
The clear preference of the 18 engineering students who participated in the study was if distance learning is necessary, the telepresence robot format is the best. They said the robot kept them more engaged and involved in the class.

Pre-Pandemic
Even before the pandemic hit, 14% of university students in the US were earning their degrees online and another 15% were using distance learning technologies some of the time. Interestingly,

the students who participated in the NSF study were split as to whether they preferred in-person or distance learning.

Key Telepresence Robot Functions

The students used the telepresence robot for three key functions: listening to the lectures, asking questions and during the break moving the robot around to talk with friends and the instructor. Telepresence robots are playing a very important role in the practice of medicine. They are now serving an important role in education even to the point of enabling Class of 2020 graduations to take place at schools like The Thunderbird School of Global Management at Arizona State University.

8.Growing Global Drone Business

Source: Volocopter

Honeywell Capitalizing on Flying Taxis & Drone Deliveries

Honeywell International, a major aerospace supplier, is launching a new autonomous aerial systems business unit. The strategic move is a bet on growth in the unmanned aviation market of flying taxis, packages and cargo delivered by unmanned vehicles and drones. Honeywell expects that business will grow into a $120 billion market by 2030. And Honeywell expects to get 20% of that market.

Technology Supplier

Honeywell's areas of expertise are building aviation electronics and autonomous flight control systems. The company doesn't build the drones. It supplies aviation electronics and autonomous flight systems to customers like Germany's Volocopter, which is building eVTOLs (electric vertical takeoff and landing) flying taxis. Another customer is UK-based Vertical Aerospace which has a prototype vehicle that can carry cargo up to 551 pounds and fly it at 50 miles per hour.

Rising Drone Demand
Honeywell has seen demand for drone deliveries dramatically increase since the pandemic began. Their aim is to greatly accelerate their electronic and autonomous systems sales into the drone, autonomous cargo delivery and flying taxi markets.

9 Japan's Ugo Avatar Robot Multi-Tasker

Source: Mira Robotics

Invented by Mira Robotics
Japan's workforce is dwindling by more than a half-million workers a year because of its aging population. To fill the worker

shortage, Japan is experiencing an acceleration in the development of new, versatile robots. A great new example is Ugo, the creation of Japanese startup company Mira Robotics.

Remote Wireless Operation

Ugo is a state-of-the-art, remote controlled Avatar robot. It is composed of a pair of height adjustable robotic arms that are mounted on wheels. It navigates with a laser on its base. And it is operated remotely through a wireless connection to a laptop and game controller.

Many Uses

Ugo has many uses. Its primary purpose is to serve as a worker robot and fill needs in Japan's dwindling workforce. It can work as a security guard, inspect equipment and clean buildings. It can also serve as a helper in homes, for instance, to fold the laundry. Given the global COVID-19 threat, the demand for robots is on the rise in order to reduce direct contact among people. Ugo has been given a hand attachment that uses ultraviolet light to kill viruses on door handles.

Robots For Rent

Mira Robotics says it takes a person 30 minutes to learn how to use Ugo. They add an operator can control 4 Ugo's at a time. Ugo can be rented for $1,000 a month. It's working in Tokyo office buildings. Ugo is part of the growing robotic workforce in our pandemic threatened world.

10.DARPA's Underwater Drones Harvesting Ocean Energy

Source: DARPA

Sustainable Energy

The concept of underwater energy harvesting by drones is being developed by the US Defense Department's Advanced Research Agency DARPA. This could not only revolutionize naval propulsion, particularly for drone submarines, by providing them a sustainable energy source. But also, if the drones transport the harvested ocean energy to ships at sea, there would be far less need to return to port to refuel. The time and energy efficiencies would be breakthrough. And the eventual commercial, consumer applications could be breakthrough.

Manta Ray Project

DARPA calls the program the Manta Ray Project. Its goal is to develop unmanned underwater vehicles - drones - that can operate underwater for very long, extended periods of time, all the while harvesting the energy of the ocean as a power source. Four companies have been selected for the job. Lockheed Martin, Northrup Grumman and Navtek LLC are working on the drones. And Metron is developing underwater energy harvesting technologies.

Developing Revolutionary Technology

At the moment, underwater drones are greatly limited by their power supplies that are often as short as 24 hours. Developing drones with greatly expanded, long range duration, along with the technology to harvest the ocean's energy, could revolutionize US Naval operations and be a tremendous game-changer for the US military. As with most of DARPA's innovations, there would undoubtedly be significant civilian applications as well.

11. Graduation 2020 Avatar Robot Ceremony

Source: Arizona State University - Dean Khagram

Virtual Graduation Ceremony with Robots

At Arizona State University, there was a graduation ceremony for the Thunderbird School of Global Management Class of 2020 graduates. But it was technologically awesome, memorable and very different. In the middle of the pandemic and lock-down in Arizona, telepresence robots, known as avatar robots, came to the rescue and saved the day. It was graduation by telepresence and a virtual graduation ceremony. May 11, 2020 was graduation day, thanks to the Dean, technology and avatar robots.

Technology to the Rescue

The telepresense robots gave students a chance to "walk in" the ceremony. The robots stood in place and were awarded each students' degree in a virtual ceremony. Live audio and video on each robot allowed the students to experience the ceremony in real-time.

Much Prep To Make This Virtual Graduation Happen

There was a lot of technology and preparations done to make this virtual event happen. Cameras pre-recorded 140 graduates as they logged in to the University from home. The grads were dressed in their caps and gowns. Each moved a remote controlled robot at the University that held an image of their face. Via the robot, they approached Dean Sanjeev Khagram to receive their hard-earned diploma and get a photo when they received it from him.

Great Effort through Technology to Honor Grads

This was not the graduation the students had been anticipating, but it certainly is an historic and memorable one. The robots are from Double Robotics of Burlingame, CA. Before the pandemic, they were primarily used to enable people to show up at weddings and other events without having to travel to them. Congratulations to the
graduates and to Dean Khagram who went the distance to honor his graduates.

12.Tokyo's Robot Hotels For COVID-19 Patients

Source: Pepper the Robot

Staffed by Robots, Doctors and Nurses

For COVID patients with mild or no symptoms, Tokyo has opened

two robot hotels to take the load off of hospitals. The hotels are staffed by robots, doctors and nurses. With a capacity to handle 2800 patient-guests, they are being used
as an innovative means to quarantine and treat COVID patients.

Pepper and Company

The Tokyo Metropolitan Government has taken charge of the two hotels. Very popular humanoid robot Pepper is at the hotels to greet the arriving guests with positive messaging: "Let's bring our hearts together and get through this." Cleaning robots are deployed to clean what the Japanese are calling "high risk red zones".

Safer Than Home

The Tokyo government feels it's safer for the mildly ill COVID patients to stay at the hotels, which are loaded with technology and medical staff to monitor them, rather than to stay at home. This is another example of robotic technology doing
a job with big exposure to the coronavirus and keeping hotel employees out of harms' way.

13.Rise of the Robotic Workforce

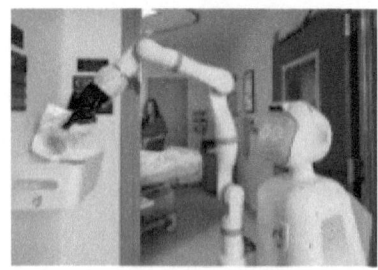

Source: Diligent Robotics' Moxi

Robots Don't Get Sick

The coronavirus pandemic is accelerating the global use of robots. The virus is so highly contagious that it is changing consumer preferences for human interaction, for instance, with a retail clerk. Dealing with a robot worker, that doesn't cough or sneeze, is a lot safer. Experts say the pandemic is rapidly expanding work opportunities for robots.

McKinsey Forecast
A forecast by the global consultancy McKinsey predicts that 1/3 of US workers will be replaced by robots by 2030. The pandemic is accelerating that trend.

Global Trends
Many companies are expanding their use of robots to foster social distancing and to cut the number of staff who have to physically come to work. For instance, Walmart is using robots to clean and sanitize their floors. In Wuhan, China a smart field hospital for COVID-19 patients was completely run by robots, controlled by medical staff in a safe, remote location. In South Korea, robots are dispensing hand sanitizers and taking temperatures. With experts saying social distancing may be necessary through 2021, the demand for health care robots is expected to dramatically rise.

Multiple Robotic Uses
Robots are being deployed for a myriad of tasks since the pandemic broke. Ultraviolet light disinfecting robots are being shipped to hospitals in Europe and China. As we start getting back to normal, we're likely to see them cleaning offices and schools. Restaurants and grocery stores are expanding their use of robots. With the priority being the health and safety of workers and customers, we're witnessing the rise of the robotic workforce. In their warehouses, Amazon and Walmart already utilize thousands of robots. They're planning on deploying thousands more.

14.Samsung's Army of Service Bots

Source: Samsung

Bot Care

Samsung recently introduced three new service robots. The Bot Care robot is for health care. It has a wide range of functionality. It's able to measure blood pressure, breathing and heart rate. Its face display allows you to make video calls. That display can also play music and provide you your daily schedule and the weather. Bot Care also provides you an alert to remind you that you haven't taken your medicine.

Bot Air

The new Bot Air contains sensors positioned around your house to monitor air quality. If the quality isn't quite right, the bot goes to the location and filters the air until the quality is fine.

Bot Retail

Bot Retail has facial recognition and vision identification to help consumers find the items they want. Customers can access store details and menus on the bot's face to purchase goods. These are three, new examples of the growing robotic service workforce.

15.Microrobotic Device Enables Jellyfish to Swim 3X Faster

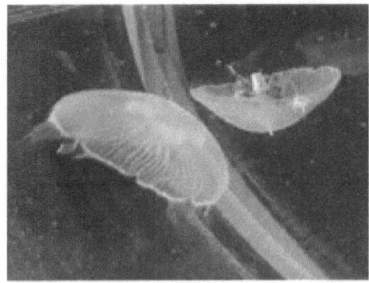

Source: Caltech

Invention with Big Potential to Monitor Health of Oceans

Scientists at Stanford University and Caltech have created a pace-maker-like device that enables natural jellyfish to swim three times their normal speed. The microelectronic, pulsing, tiny, prosthetic, micro-robotic device generates electric jolts that make the fish triple their speed while using just twice as much energy. The scientists and engineers involved in the project say the device causes no additional stress to the sea creatures. They say the potential for this system is as vast as the ocean itself.

Network of Live Ocean Monitors

The scientific team says the combination of speed and energy efficiency from this device opens the possibility of using jellyfish to gather data from across the world's oceans. They envision equipping the jellyfish with sensors to track ocean temperatures, salinity and oxygen levels. They say this could lead to a truly global network of ocean monitors composed of robotic jellyfish that would cost just a few dollars to instrument. The fish would generate energy from their normal ocean food. Obviously this system needs a tremendous amount of research and development. But, it's a fascinating look at instrumented jellyfish monitoring the health of world oceans.

16.Impressive Bio-Inspired Robots

Source: Purdue University

Bioinspired by Chameleons, Salamanders and Toads

Researchers at Purdue University have invented a new class of high speed, soft robots and actuators. They are bio-inspired by the high speed, elastic motions of salamanders, chameleons and toads that launch their tongues in a tenth of a second to catch an insect. Purdue's new soft robot can catch a live, flying beetle in the same way, in a blink of an eye. This robotic innovation is big for potential use in accelerated automation.

New Class of Soft Robots

The Purdue scientists' new innovation is a new class of entirely soft robots and actuators. They are capable of recreating bioinspired high powered, high speed motion using stored elastic energy. They're composed of stretchable polymers like a rubber band with channels that expand with pressurization.

Amazing Robotics

This robot expands five times in length and can catch a live flying beetle in 120 milliseconds. The robot's elastic energy is stored by stretching its body during fabrication. The scientists believe such high speed robotics could achieve automated tasks more accurately and much faster.

17.Smart Robodog Learns Like a Dog

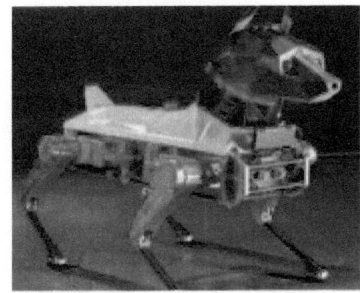

Source: Florida Atlantic University

New Invention from Florida Atlantic University Roboticists

Astro is a deep learning and artificial intelligence powered, quadruped robodog created by a team of roboticists and other experts at Florida Atlantic University. Like your dog, it learns by trial and error. It has a deep neural network that enables it to learn and perform tricks. Astro's 3D printed head is designed to resemble a Doberman and has a computerized brain. The team calls it "a puppy in training" but it can sit, stand and lie down. The robot can see and hear.

Loaded with Technology

Astro contains cameras, high tech radar sensors, radar imaging and a directional microphone. Most importantly, it has four teraflops of processing power or the capacity for 4 trillion computations per second with its Nvidia Jetson TX2 graphics processing units. All of this technology allows it to be trained like a dog. The FAU team says Astro will be able to respond to hand signals, identify colors, understand different languages, coordinate with drones, recognize faces and a lot more.

Future for Astro

Astro is a "puppy" learning new commands but this is not just fun and games. For Astro, future functions include bomb detection, search and rescue, work as a service dog, medical diagnosis monitoring and potentially even making real-time decisions based on experience.

18.Robots in COVID-19 Fight

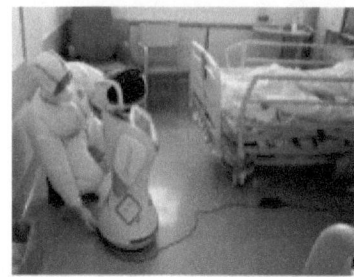

Source: CloudMinds

Caregiving Robots on the Frontlines
With the insidious contagion of the coronavirus pandemic, robots are being increasingly used on the frontlines
of COVID 19 patient care. Caregiving robots are fast,
efficient and "contagion proof". With medical staff supervising remotely, they keep doctors and nurses out of harms' way.

Wuhan Smart Field Hospital
Wuhan, China is where the global pandemic started.
In a smart field hospital, they temporarily used a team of
robots to care for COVID 19 patients. The robots communicated with the patients, served them meals and took their temperatures. The robots were managed by
the medical team remotely. The patients also wore wristbands that monitored their blood pressure and other vitals. The patients remained in the field hospital for just a few days.

Future of Medicine

This field hospital in China is a glimpse at the future of medicine during global pandemics. Robots directly in contact with infected patients while health care providers supervise and manage from safe distances. The robots deployed were
made by CloudMinds, with operations in California and Beijing. The company says the robots "completely ran the field hospital". No health care workers were exposed but still in control of the robotic team.

Global Pattern
Because of the coronavirus, the use of medical robots is on the rise. In some Singapore hospitals, robots are delivering meals and medication to COVID-19 patients. Hospital patients in Israel, Thailand and elsewhere meet with robots to consult with their doctors via live videoconferencing. Robotic machines are being used to scan for viruses, location to location, and to disinfect facilities from viruses and bacteria. It's a new
world. And robots are a key. They don't cough and sneeze.

19.Robot Satellite Repairmen

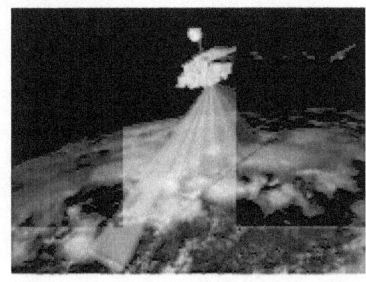

Source: NASA

New DARPA Space Initiative
The US Department of Defense Advanced Research Projects Agency, DARPA, has launched an innovative program to extend

the life of US satellites orbiting in space. In essence, DARPA is creating a robotic system to inspect, service, repair, improve and extend the lives of US satellites in orbit. The program promises significant money savings and much longer and efficient use of satellites.

Space Logistics

DARPA is partnering with Space Logistics, a subsidiary of Northrup Grumman, to develop and deploy advanced robotic capabilities in space. The first step is to develop what DARPA is calling a "DEXTEROUS ROBOTIC SERVICE". The technology needed would increase satellite life spans, improve their reliability and resilience. All of which would save the US government and space companies significant money lost when a satellite's operations breakdown and are unfixable in their space orbit.

In GEO

The robotic servicing system would be in GEO or geosynchronous orbit, as are the satellites that it will be servicing. Space Logistics will build the robotic servicing spacecraft, integrate it with a launch vehicle, do the launch and then offer commercial satellite servicing to government and commercial client spacecraft. This promises to be a much needed robotic maintenance and repair service for satellites.

20.Europe's Robot Airport Parking

Source: Stanley Robotics

Deployed and Being Expanded at Top European Airport

The outdoor, robotic parking service developed by Stanley Robotics tested so well it has been expanded at Lyon Airport in France. The Airport is expanding the service from 500 parking spaces to 2,000 by late 2020 and 6,000 beyond that. Lyon Airport and Stanley Robotics have invented a very innovative and convenient airport parking lot that is leading the way in passenger parking convenience by robotics.

Convenient and Easy Airport Parking by Robot Valets

The expanded robotic service is run by seven autonomous robots working simultaneously with 28 cabins to drop off and pick up cars. Lyon Airport CEO Tanguy Bertolus says the service enhances customer convenience and the airport experience and cuts the environmental impact of airport activities. In fact, it cuts the carbon dioxide emissions from cars looking for a parking space significantly. Lyon was named Europe's best airport in 2019 in the 10 million to 25 million passenger category. Airport officials believe that the Stanley Robotics system significantly contributed to their win.

Robotics Economy

Stanley Robotics is part of the robotics future in service that is underway. Their parking service is very easy to use. Passengers book their parking space on the Airport website, drop their car off at the designated cabin, go to their plane using nearby shuttle buses. When they return from their trip, they pick up their car at a designated cabin. It's a robotic answer to tough parking at global airports.

21.Misty 11 - Robot Concierge

Source: Misty Robotics

Your Robotic Hotel Experience

Misty 11 has great aspirations to be a key part of the hospitality industry. The cute, little rolling robot is the invention of Misty Robotics, based in Boulder, Colorado. Misty can detect humans, greet them, interact, provide information and responses. She can make reservations, call for assistance and even be part of point of sales systems. The little robot gives new meaning to the hospitality industry. Her inventors have designed her to go to work in a hotel near you very soon.

Open Source Code

What's ingenious about Misty is she is not limited to one task. This is part of an important, new emerging trend to design robots that are not limited to one use. They have capabilities that can be used in many ways. In the case of Misty, from room mapping to greeting customers and taking reservations. The open source code allows developers to customize her for a myriad of duties. The inventors at Misty Robotics say that Misty as a concierge is a "starting point to accelerate" her deployment in the hospitality industry.

22.Ford and Digit the Robot

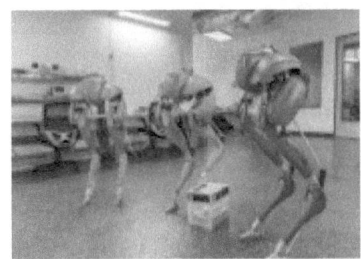

Source: Agility Robotics

Ford Purchases Digits

Digit is a 5 foot tall, 2-legged robot invented by Agility Robotics. Digit walked around the 2020 CES trade show in Las Vegas at its debut taking bows. It should be proud of itself. Ford became the first customer to purchase Digits and it has big plans for them.

AV + Digit Plans

Ford says the initial plan is to deploy Digits for research and development. But the goal is to employ Digits as part of a new package-delivery service driven by Ford autonomous vehicles. Digits would be on board to deliver the package to the owner's door.

Agility Robotics

Agility Robotics was founded in 2015 and is headquartered in Albany, Oregon. It makes advanced bi-pedal robots with human-like capabilities for diverse markets including logistics/delivery, telepresence and automated inspections. Agility Robotics designs robots that can operate in "human spaces".

23.Harvard's RoboBee Breakthrough

Source: Harvard University

Search & Rescue Missions
Harvard's robobees are the first microbots powered by soft actuators to achieve controlled flight. They have soft artificial muscles that enable them to survive crashes and collisions making them perfect for search and rescue missions in dangerous, cluttered environments. The robobees are so sturdy, dexterous and resilient they can even crash into a wall or collide with another robobee without any damage.

Hoverbots
The tiny robots are equipped with actuators made from dielectric elastomers that deform when hit with an electrical current. The actuators are soft and the Harvard team says they're easily assembled and scaled up. Unlike other drones made with soft actuators, the robobees have enough power density to hover in place.

Going for Commercialization
The Harvard team has created a number of models including one with 8 wings and 4 actuators that can do controlled hovering flight, which is a first. They feel the sky is the limit for the number of robots of this type that they can build. Harvard's Office of Technology Development has protected the intellectual property of this invention and is exploring commercialization.

24.Safer Working Robots Through Sensitive Skin

Source: Technical University of Munich

Touch Sensitive Artificial Skin Makes for Safer Robots
This is new innovation from the Technical University of Munich. Their H-1 autonomous, humanoid robot is covered with touch sensitive, synthetic artificial skin. The skin is composed of 13,000 separate sensors. The sensors enable the robot to feel the touch of humans. New algorithms made it possible to apply artificial skin to a human sized robot.

Safer Working Robots
This is an important development because of concerns about humans getting hurt working side by side with robots. Biologically-inspired artificial skin improves the robot's sensory ability, making it possible for the robot to sense its own body and surroundings.

Senses Like the Human Brain
The robot's artificial skin is able to measure temperature, pres-

sure, proximity and acceleration. The sensors are an event-based system that transmits information only when a value is changed. It works in a manner similar to the human nervous system and reduces the computer processing demand by 90%. The artificial skin helps the robot to operate more safely when near people and gives it the ability to anticipate and actively avoid accidents.

25.Tunabot – New Robot, New Underwater Propulsion

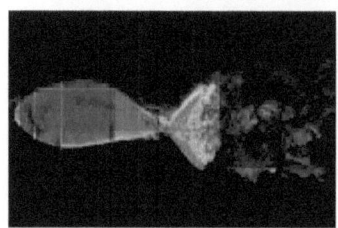

Source: University of Virginia

Prototype for Next G Underwater Vehicle Propulsion
The tuna is one of the fastest fish in the sea. A mechanical engineering team from University of Virginia, in collaboration with biologists from Harvard University, have invented a robofish that can swim as fast as a yellowfin tuna. The team says it's not about the robot. It's about inventing a new, faster and more efficient underwater propulsion system for manned and unmanned underwater vehicles.

Physics of Fish Propulsion
At Harvard and UVA, Tunabot is tethered in a large flow tank with a green laser to measure fluid motion as it swims. Yellowfin tuna grow to 7 feet. The Tunabot is 10 inches long. The purpose of this research is to better understand the physics of fish propulsion to develop the next generation of underwater vehicles with fish like propulsion systems. The team says the ultimate goal "is to surpass biology" with new, quicker and more efficient propulsion.

According to the researchers, Tunabot is the world's fastest robot fish.

Surpassing Biology
The device has been tested at both Harvard and UVA. It can swim with a maximum speed equal to the real fish, which is 4 body lengths per second. Tunabot's body is flexible and it swims just like a tuna. With a 10 watt battery pack, it can go 1.3 feet per second for a distance of 5.6 miles. If the speed is increased to 3.3 feet per second, the range is 2.5 miles. This is just the beginning for this new technology. It could possibly be used for underwater surveillance, as part of the funding is from the US Office of Naval Research. But the aim is beyond that: a brand new, innovative propulsion system that is bio-inspired.

26.Toyota's Robot For Your House

Source: Toyota Robot

Home Support Robots
Toyota has big plans for service robots for your home. To accelerate R&D, Toyota joined forces on research with Tokyo based Preferred Networks, which specializes in AI. Japan's largest automaker wants the robots to be capable of learning "in a typical living environment".

Robots in Homes and Hospitals
Toyota and Preferred Networks have been collaborating on

driverless vehicles since 2014. In the dramatically changing automotive environment, Toyota has been pushing forward with new technologies such as electric, hydrogen and autonomous cars. The company views robots as a big part of the future. Not only does Toyota want to put robots in your home but also in hospitals.

Home Support Robots

The arrangement with Preferred Networks is a three year deal that includes sharing of intellectual property. The focus is on Toyota's most promising robotic system - the Home Support Robot (HSR). Toyota had planned to put HSRs to work at the 2020 Summer Olympics in Tokyo for simple duties like delivering drinks to spectators. Because of COVID-19, the Games were cancelled.

27. Robots to Install Telescopes on the Moon

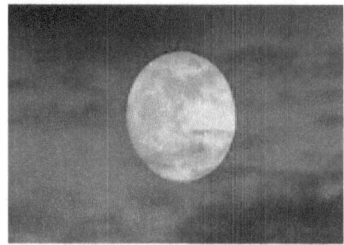

Source: Stock image of Moon

Telescopes to Look Deep Into the Universe

A NASA funded lab at the University of Colorado is developing robots to deploy small, highly advanced telescopes on the far side of the Moon. In the next few years, the NASA team will send a rover aboard a lunar lander spacecraft and place it on the dark, far side of the Moon.

Humans & Machines Working Together
The plan is to set-up a network of small telescopes. The deployment will be performed by the rover's robotic arm. The arm will be controlled by astronauts in an orbiting lunar station called Gateway. Gateway will serve as transit to and from the Moon and as a refueling station for deep space missions.

New Telescopic Views of Deep Space
On the far side of the Moon, the telescope will be free of light and noise. That will enable a unique and pristine gaze into the deep reaches of space. It's a leading edge example of projects underway by NASA, private companies like Space X and other nations. These missions will open up and change the landscape of the Moon forever.

28.Jellyfishbot

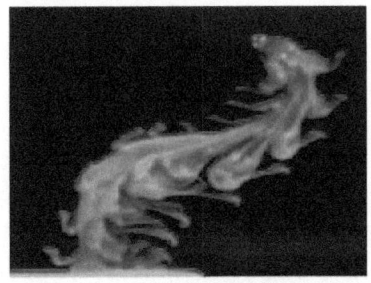

Source: Max Planck Institute

Big Potential for Medical and Environmental Applications
Scientists at the Max Planck Institute in Germany have created a tiny soft, swimming robot inspired by jellyfish. Just a few millimeters in size, it can perform a variety of tasks. Besides swimming, it can transport objects, burrow, mix different fluids and

leave a chemical trail behind it.

Magnetic Force

The robot is composed of a magnetic core connected to 8 bendable flaps that "swim" in response to the application of an oscillating magnetic field. This type of robot has great potential for targeted medical treatments. One example would be the tiny robot gliding through the human body to locate cancerous tissue and then treating it with targeted, timed medication. And, the researchers say it can be used as a model to help understand how the changing environment is impacting jellyfish.

29.New Boxer Robot That Delivers

Source: CMC CartonWrap

Superstar Customers

The Italian packaging company CMC in the medieval town of Citta di Castello has robot machines filling some very big orders for big clients and it delivers. The robotic packaging machine is called CartonWrap. It can box and seal 600 to 1000 items of every shape and size every hour, 24/7, including in tailor made boxes. Clients of the automated packaging machines include Amazon, Walmart and Gucci.

Big Business in Italy

CMC is one of 630 companies in Italy making automated packaging machines. This is the new robotic, automated economy and it is one of Italy's fastest growing industries. In fact, the sector is accelerating 9X's faster than the overall Italian economy. It's another example of robotic technology cutting time, costs and in this case packaging waste and creating a whole new avenue of global business that's a disruptor.

30.Robot That Actually Drives the Car

Source: IVObility.

The Bot's in the Driver's Seat

Israel start-up IVObility is developing a robot that can sit in the driver's seat of an ordinary car and drive it. This gives new meaning to autonomous driving. The robot is an independent driver. There are no wires connecting it to the vehicle. The robot drives like a human by using sensors and cameras that see exactly what a human driver sees and the bot drives accordingly. This is a new and fascinating approach to autonomous vehicles.

Accelerating Self-Driving Innovation

The driver robot will hit the market at first for government and commercial uses that don't run on public roads. A consumer version is also being considered. What's remarkable is that the robot in the driver's seat doesn't require special drive-by-wire equipment. It's able to drive autonomously on its own technological capabilities.

31.Ai-Da's Robot Artist

Source: Aiden Meller's Ai-Da

Already a Sell-out
Ai-Da has been described by her British inventor as the world's most realistic, humanoid robot artist. She has had a solo exhibit of 8 drawings, 20 paintings, 4 sculptures and several videos. The inventor is Aiden Meller who also owns an art gallery. He says that Ai-Da brings a new technological voice to the world of art. In fact, Ai-Da is a work of technological art.

AI, Technology and Art
The robot can paint from sight with cameras in her eyes. AI algorithms, created by scientists at the University of Oxford, enable her to paint. She sketches with pen or pencil and then the sketch is printed onto canvas. A human artist paints over the sketch. Her solo show at St. John College was sold out with over a million pounds of artwork sold.

32.Robotic Furniture Transforming Living Spaces

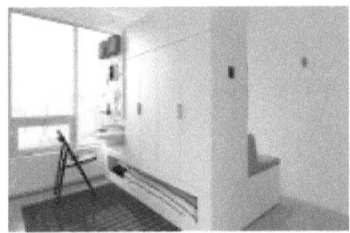

Source: IKEA Rognan

Making the Most of Small Living Spaces
The world's biggest furniture brand, Sweden's IKEA, is teaming up with a US startup Ori that began at MIT to launch shape changing, multifunctional robotic furniture. They've unveiled one of the top new innovations of the year - a moving, robotic room divider that makes the most of small living spaces, creating a living room, sleeping area, walk-in closet and more.

Ikea Rognan
This is the next generation of furniture where robotics, technology and design come together to address small living spaces. The Ikea Rognan is an L shaped storage unit that moves with motors and a switch. The functionality is impressive innovation. Rognan serves as a movable room divider. On one side there's a full bed that can be pulled out and secured back into the unit. On the other side, there's a sofa to create a living room. And on both sides of the unit, you can customize shelves and drawers for storage.

Launching 2020
Ikea has been developing this breakthrough piece of robotic furniture with Ori for two years and has now licensed it for its stores. Pricing hasn't been announced yet but Ikea is known to come in on the more affordable side. Rognan will launch first in Hong Kong and Japan.

33. Door to Door Robotics

Source: Amoeba Energy AE-01

AE-01 to Deliver Packages to Your Door

A team at Amoeba Energy in Japan has created a soft treaded robot that's able to climb any type of stairs. The tracks on the robot are made of a soft material (EPDM) that can change form to climb stairs. The robot is covered in foam to make sure it doesn't hurt humans. It's designed to carry and deliver packages and other loads like luggage.

Amoeba Inspired

The soft bodied robot was inspired by amoebas that deform their soft bodies to survive and adjust to various environments. The robot, called AE-01, can deform, climb and carry loads. Amoeba Energy says soft-bodied robots are reshaping the future of mobility. AE-01 has tested very well. The company is developing a commercial version. They hope to put it into package delivery service and as a helper in your home to do the heavy lifting.

34. Italy's Very Strong Robot

Source: IIT - Robot Towing Passenger Plane

HyQReal

The new robot HyQ Real is autonomously powered and is so strong, it can tow a 3 ton passenger plane. It did so at an airport in Italy, towing it with ease for 33 feet. It's a robot dog, a quadruped developed by researchers at the Istituto Italiano di Tecnologic (IIT).

Robodog

They've designed the very strong robodog to help humans in emergency situations. The robot weighs 280 pounds and is about 4 feet long. It has very strong traction on the ground thanks to its custom made feet with special rubber grips. The researchers have been developing HyQReal since 2007. The plane towing robot is number 7 in the series.

35.Robot Helps You Lift Objects

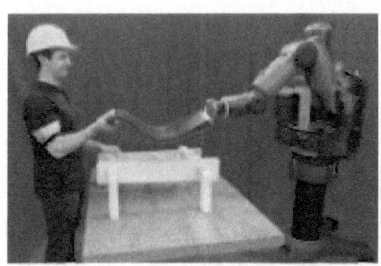

Source: MIT

Sharing the Load with a Robot

MIT's robot helps you lift heavy objects by taking signals from your biceps and triceps. It's a case of a human and a robot getting a heavy lifting job done together. MIT roboticists think this approach is a key part of the workplace of the future.....truly teaming humans with robots for certain tasks.

AI Analyzing 80 Times Per Second

The robot works by directly reading the electrical signals produced by a person's biceps. Here's how the technology works. Electrodes are placed on the person's upper arm and connected with wires to the robot. The robot reads how the person is moving and then gives them the help they need to lift the object. To do so, the robot's Artificial Intelligence is analyzing what it sees 80 times per second. It's a cutting edge example of humans and machines working together to get a job done more quickly and easily.

36.New Hummingbird Drone

Source: Stock Image

Hovers and Turns Like a Hummingbird

Purdue University researchers have developed a drone that has been taught to maneuver like a hummingbird. It hovers and makes quick turns in flight. It is important new technology for difficult to get to places in search and rescue and surveillance. It can nimbly travel through the air, debris and crowded spaces. It's enabled by artificial intelligence.

Machine Learning

The robotic hummingbird was trained on machine learning algorithms based on hummingbird flying techniques. The robot is able to develop maps as it tracks and touches surfaces and doesn't need a camera. The little robots have 3D printed bodies, wings made of carbon fibers and laser cut membranes. They range in weight from 12 grams which is the average weight of a male, hummingbird to 1 gram, which is insect-sized. The team expects to add sensing technology like cameras and GPS to their new robots.

37.China's Robot Ship

Source: China's Robot Ship

Launches Sounding Rockets to Monitor Atmosphere

3/4 of the earth's surface is ocean and studying the atmosphere over them is very difficult and expensive by ships and planes.

Advanced Weather and Atmospheric Monitor

China has built the world's first robotic, partially submersible ship to launch small, sounding rockets to study the atmosphere and weather systems over oceans to provide more accurate forecasts.

Typhoon Patrol

Called an unmanned, semisubmersible vehicle, the boat is designed to sail into bad weather, deploy sounding rockets and gather critical data from the meteorological equipment it carries. The rockets do brief flights through different regions of the atmosphere up to five miles. The system has tested well. China wants to deploy a network of the boats to study typhoons. They also want to equip the boats with advanced sensors to look below the ocean surface as well as above.

38. Jenga Playing Robot

Courtesy: MIT

MIT Robot Has Interactive Perception & Manipulation

MIT engineers are on a robotic roll. They've created a robot that combines vision and touch to remove Jenga blocks without toppling the tower of the game. This is being called a new kind of robot using interactive perception and sophisticated manipulation that experts are calling remarkable robotics.

Robot Well Equipped

The robot has a gripper that is soft pronged. It also is equipped with an external camera and a force-sensing wrist cuff - all of which enables it to play "a mean game of Jenga." In Jenga, players take turns at taking out a block from a tower of 54 blocks and then place the block on top, making it more unstable. Whoever causes it to topple loses.

Remarkable Robotics' Machine Learning

The robot uses machine learning to select the perfect block to take out. It uses feedback from the camera and cuffs to compare measurements to move the block out perfectly. This is viewed as remarkable robotics because to play the game the robot requires a complex understanding of forces of physics like spatial arrangement, push and pull. There are many applications of this new technology including smart phone assembly lines.

39. AIBO Pet Robot

Source: Sony's AIBO

Sony's High Tech Pet
Sony launched a significantly upgraded version of their puppy bot AIDO, giving it a new lease on life. Sony says the new hardware and AI perfectly come together making it the most realistic and advanced robot dog they've ever created.

Cute with Personality
This little robot has personality and mimics a real pet dog. It has 4,000 parts and 22 actuators that bring it to life. There's a camera in its nose and one in front of the tail that enables it to identify members of the household and map the home environment. It relies on sophisticated sensors and AI to maneuver, similar to what's in self-driving cars. The eyes are OLED screens. It barks and wags its tail when petted. The cameras guide the puppy to its charging station when it needs a recharge. Priced at $1700, the puppy romps and plays for two hours on a charge.

40.China's New Bot Delivery System

Source: Segway-Ninebot's Loomo

Loomo

From San Francisco to Beijing, armies of service bots for package deliveries are being developed. Beijing based Segway-Ninebot is developing delivery robots. Its new one is Loomo and is expected to hit the market this year. It will be making deliveries to shopping malls and office buildings. It looks like an office file cabinet on wheels and has been developed from a hoverboard into a robot. Here are a few more leading examples.

German-Swiss ANYmal

ANYmal is a 4 legged robot dog that can walk packages right up to your door. It was created by German automotive company Continental and Swiss robotics company ANYbotics.

Serve

Another delivery bot being developed is Serve by San Francisco based delivery company Postmates. It started work in Los Angeles with more cities to follow.

Pepsi

Pepsi deployed its Snackbots at the University of Pacific before the pandemic and lockdowns hit. It delivers snacks to students who need to meet the bot at a designated destination. Robo delivery services are becoming a big part of the global robotic industry.

41.Hyundai's Concept Car Climbs Stairs

Source: Hyundai Elevate Concept Car

The Elevate Gives New Meaning to "Door to Door"
It's walking automotive innovation. Hyundai has a very impressive concept car, The Elevate.

Robotic Car
It drives on a traditional suspension. But each wheel is connected to a robotic leg that can be deployed to walk over rough terrain and take you where you can't drive.

Humanitarian Concept
This robotic car can climb a 5 foot vertical wall, go up and down stairs and walk. The idea is to help first responders and people with disabilities traverse "the last mile or last 100 yards". Hyundai at this time doesn't have plans to put it into full production. But it is a great piece of robotic innovation.

42.Deliveries By Robodogs

Source: ANYmal Delivery Robot

Robots Delivering
It's a 4 legged delivery robot that can walk packages right to your front door. And, it's an ingenious way to automate the delivery of food and packages, especially in urban areas. It's a new robotic and autonomous car delivery system developed by the Swiss robotic company ANYbotics in collaboration with the German automotive company Continental.

High Five Paws
ANYmal weighs 65 pounds and can carry up to 22 pounds. The delivery robot was demonstrated at the Consumer Electronics Show. It hopped out of Continental's autonomous vehicle CUbE. The robodog jumped over a scooter, walked up a flight of stairs, used a paw to ring the doorbell and placed a package on the door-step. It then did a little celebratory dance.

Very High Tech Tease
The robodog can move a meter in a second. It has wide angle cam-eras, sensors in its feet and advanced LIDAR to map its location

and navigate. Continental says it wanted to "tease" the delivery industry with the system. It certainly has and is also a bit of a tease about when it is going to take the system to market.

43.Ford's New Robot Survival

Ford's Survival Robot

Helps Workers & a Self-Learner

Ford engineers have developed "Survival". They named their robot Survival because of its ability to learn its way around a factory floor on its own. This is a self-driving robot that delivers spare parts where needed, dodges objects and changes its route and stops if necessary. Ford has deployed it in Europe. Ford itself says their employees first thought they were part of a sci-fi movie. But now Survival is part of the process and works perfectly around them. The robot frees-up humans for more important work.

No Set-Up Needed

Survival eliminates the need for any set-up assistance. It notes and remaps objects in its route and adapts to its environment. It's programmed to learn the whole factory floor and with sensors navigate on its own. Ford engineers say this robot can save 40 employee hours daily by delivering parts and welding materials. It's also about what self-driving vehicles are possible of including for cars.

44.First in Medical Robotics

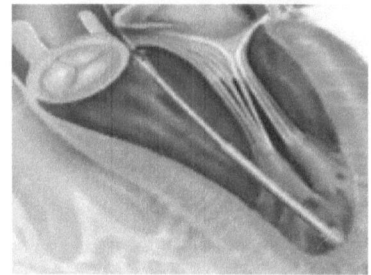

Source: Boston Children's Hospital

Robotic Catheter Noninvasively Repairs Leaking Hearts
Boston Children's Hospital bioengineers have invented an autonomous robot that is able to autonomously navigate inside the body. That is a medical robotic first. It's a robotic catheter and offers a non-invasive way to repair leaking heart valves.

Artificial Intelligence Enabled
The device has an optical sensor powered by AI and image processing algorithms that guide the device through the heart. The robot uses a "wall following" navigational technique to determine where it is and where it needs to go.

Breakthrough Medical Innovation
Once it arrives at the leaking valve, the surgeon takes over to remotely operate a tool on the device and repair the leak. This breakthrough in robotic medicine worked a number of times on animal models. This is the world's first demonstration of a robot that can autonomously navigate in the body. It may usher in a world of new possibilities in the practice of medicine.

45.Soft Knitted Robots

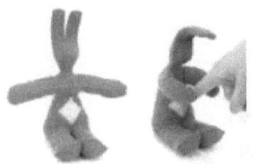

Soft-Actuated Robots From Commercial Knitting Machines

This is a new approach to exploring the potential of soft robotics. Researchers at Carnegie Mellon University (CMU) have developed a soft robotic toy through knitting. The robots are actuated by tendons attached to motors. It's a unique way to develop soft robotics which is a growing field. The goal is to make robots inherently safe around people.

Interactive Objects

The researchers say their approach may be used to make cost effective soft robots and wearable technologies. They believe the many soft objects that surround us - like sweaters - may become interactive with their technology.

New Technological Approach

The CMU team used specially created software and computationally controlled knitting machines to create a plush toy with the tendons embedded in it. The tendons are attached to motors, allowing the robot to move when triggered. For instance, the robot gives a hug when poked in the stomach. According to the researchers, actuated soft components would be cheap to produce on commercial knitting machines. The National Science Foundation funded part of the research.

46.Berkeley's New Robotic Arm

Source: UC Berkeley

Artificial Intelligence and Robots
It's a new, low cost robotic arm developed by researchers at the University of California at Berkeley. They've designed it to be an inexpensive but powerful platform for Artificial Intelligence and robot experimentation.

AI Enabled Robots
The purpose is to advance AI-enabled robots that can easily adapt to changes in their work environment. Most industrial robots are powerful and precise through meticulous programming and don't easily adapt to changing circumstances.

New Technology Revolution
The team at Berkeley thinks that robotic platforms for AI and robot experimentation will usher in a new technological revolution.

47.Squishy Robots for Disasters

Source: Squishy Robotics

For Search and Rescue Missions

Round, squishy robots capable of being dropped 600 feet and surviving have been developed by engineers at UC Berkeley and Squishy Robotics, a startup company founded by UC Berkeley Engineering Professor Dr. Alice Agogino. The robots are designed for use in disaster areas to collect information on the ground.

Space Bots

These "tensegrity" robots were originally designed to explore Saturn's moon Titan. They would be dropped from a spacecraft onto Titan. The UC Berkeley team realized they could be used in disaster areas on earth by equipping them with sensors able to detect conditions like dangerous gases. The robots relay that information to first responders.

Rapid Deployment

The sensor robots can be rapidly deployed, are mobile and, according to Dr. Agogino can work as co-robots with their human partners at disaster scenes.

48.Tiny Robotic Teeth Advances

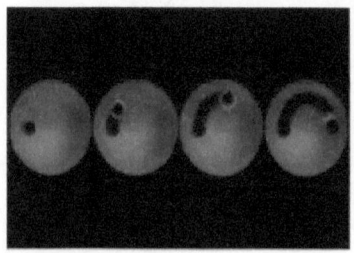

Source: UPENN's Microrobot Teeth Cleaners

New Innovation from the University of Pennsylvania
A team of engineers, dentists and biologists at the University of Pennsylvania have developed a crew of microscopic, robotic cleaners for your teeth. The new technology combines two robotic systems: one for the surface of the teeth and the other for inside, tough to get to places between the teeth and inside the teeth such as for root canals. Both robotic systems rely on iron-oxide nanoparticles.

CARs
The UPENN researchers call their system CARs, or Catalytic Anti-microbial Robots. The microbots can be steered magnetically to breakdown biofilm and accumulated material like plaque on teeth. This is a robotic biofilm removal system that has a wide range of applications. The team continues to work on their technology. They believe other applications include keeping water pipes and catheters clean as well as reducing tooth decay, endodontic infections and teeth implant contamination.

49.KIKI, ADORABLE KIKI

Zoetic AI's KiKi

New World of Robo-Pets

KiKi is a little, robot companion that's a bit of a genius. It learns from its environment through touch, sight and hearing. The company is further developing this petbot but there's a lot of interesting tech in it already. KiKi is enabled by Artificial Intelligence to learn from and interact with its owner and environment.

Very HighTech Pet

There's a camera in the nose to learn and recognize the owner. As you move around the room, KiKi's eyes are faithfully watching you like your pet dog or cat. And, just like your pet, KiKi develops a personality based on interaction with you. KiKi will be as much as a friend to you as you are to it. KiKi has 12 different touch points. Its batteries last 2 hours. Zoetic AI is developing more versions of KiKi including a dog robo companion.

50.UBTECH's NEW ROBOTIC/AI INNOVATION

Source: UBTECH's Walker

Walker and Cruzr

UBTECH unveiled a pair of new, advanced robots. They are new versions of the company's Walker bipedal humanoid robot and Cruzr. Cruzr is a cloud-based, intelligent service robot. The company says its performance and reliability are greatly improved. Walker, as you can see in the photo, has arms and hands and seems to enjoy piano music. UBTECH is a leading, global AI and humanoid robotics company based in China. The new robots are enabled by advanced artificial intelligence and robotic technology.

Retail Cruzr

Source: UBTECH's Cruzr

Customers' Service Bot

UBTECH says the robots are specifically designed to enhance the home and retail environments. Cruzr has servo motors, a powerful processor and a redesigned audio chamber for better sound quality. It's started working in retail stores in Europe to help customers. It's available for sale globally.

Walker the Home Bot

Walker is a big "agile companion robot" that now can grasp and manipulate objects like soda cans and bottles. It is 4.75 feet tall and weights 170 pounds. It has 36 actuators and a full range of sensory systems. The robot is equipped with smart home control and face and object recognition. It also has multi-mode interaction with vision, voice and touch, making it a perceptive, interactive home assistant.

51. World's First Ice Skating Robot

Source: ETH Zurich

Plays Ice Hockey Too - Ice Skating Machine

Roboticists at ETH Zurich have created the world's first ice skating robot. It can even play ice hockey with humans. Its name is Skaterbot and it is an incredible learner.

Moves on the Ice

Lead researcher Prof. Stelian Coros of the Computational Robotics Lab at ETH says the only thing his team did was to inform the robot how one skate worked and behaved on ice. They then told Skaterbot it was free to move in the direction of the blades. The robot took that information and figured out on its own how to move and skate on ice.

Quadruped

This is a modular, quadruped robot. Based on the information provided by the team, the robot determines where its legs need to be to stay balanced on the ice. The researchers believe their robot has many potential uses for deliveries, helping people and search and rescue. It's remarkable innovation - a skating machine that performs very well as an ice skater and ice hockey player.

52.China's Restaurant Robots

Source: Haidilao's Restaurant in Beijing

Robotic Restaurants

China's hottest restaurant chain by sales Haidilao is rolling out robot staffed and operated restaurants. The chain has more than 360 restaurants globally including in the US, Japan and Taiwan. It opened Beijing's first robot staffed restaurant and looks to spread the concept across its system, starting first in China.

Partnership with Panasonic

Haidilao is partnering with Panasonic to automate its restaurants. It's deploying robots taking orders, preparing and delivering meals, from a fully automated kitchen. The billionaire founder Zhang Yong says the technology is allowing him to run more efficient, lower cost restaurants. His goal is to utilize the cost savings and efficiencies to roll out 5,000 restaurants globally. This is a megatrend in the global restaurant industry. The question is: should you tip your robot waiter? That's totally up to you.

53.Revolutionary Robotic Arm

Source: Franka Emika Robotics

The Panda
Germany based robotics firm Franka Emika has built a break-through robotic arm specifically geared for small business. It's called The Panda, is easy to program and is priced at $11,000. It was cited by Time as one of the top inventions of the past several years.

Lightweight Robotic System
The Panda can move in 7 axes. The company says it was designed to mimic human agility and a sense of touch. It can be programmed to perform specific tasks and is interconnected and adaptive. Panda can build circuit boards, conduct scientific experiments and pre-test equipment. Two Pandas can work together and build a third Panda.

Next Generation
It's expected that a redesigned version will be developed for use at home as a helping hand, for instance, to chop vegetables or to assist elderly residents.

54.World's 1st AI Powered Pet Feeder

Source: Volta and Pet Electronics of New York

Innovation Machine that Caters to Your Pet

As a dog lover, I'm fascinated by this innovation for my German Shorthaired Pointer Rudi. He's just 1 years old, athletic and thin but a foodie. The new innovation is called Mookkie and it was invented by the Italian Tech company Volta. It's a pet feeder with AI intelligence that recognizes each of your pets' faces and feeds them accordingly. My other pet is Biddie, a 6 year old Smooth Fox Terrier who is another foodie and a terrier about it. This AI pet bowl feeder received the CES Innovation Award in the Smart Homes Category.

Right Portions

This device gives each pet the right amount of food. It works for dogs, cats and pets with special dietary conditions and needs. It's an innovation standout. It's priced at $189. It also sends you an alert on your cell phone that the bowl is empty and you need to refill. Interesting innovative tech for pet lovers. Volta developed this AI driven pet bowl with Pet Electronics of New York.

55.Very Smart Microrobots

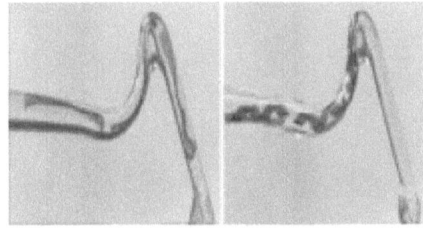

Source: EPFL

Breakthrough Robotic Innovation from Switzerland

Scientists at the Ecole Polytechnique Federale de Lausanne (EPFL) and ETH Zurich have developed very smart, highly flexible microrobots. They can change shape depending on their surroundings
.

For Targeted Drug Therapy Delivery

The tiny robots are modeled on microorganisms like bacteria that change shape as their surrounding conditions change. They are fully biocompatible. The bots optimize their movements in order to get to hard to reach places in the human body. They're thought to have the potential to revolutionize targeted drug delivery.

Nanocomposites and Nanoparticles

The bots are made of hydrogel nanocomposites. They contain magnetic nanoparticles allowing them to be controlled by an electromagnetic field. They are so tiny and flexible they're able to swim through narrow blood vessels, arrive at the target and deliver drug therapies. This is revolutionary robotics that could revolutionize medicine. Their research was reported in the journal Science Advances.

"NEW ROBOTS FOR THE 2020's" © BY EDWARD KANE ON AMAZON

Copyright 2020 © Edward Kane

ALL RIGHTS RESERVED

BOOKS BY
JOURNALIST
EDWARD KANE

Non-fiction books on innovations and discoveries across industries in 2020 and 2019.

SMART DEVICES FOR THE 2020's
Just published by Amazon and Kindle amazon.com/author/ekane

FUTURE TRAVEL VEHICLES

Just published by Amazon and Kindle amazon.com/author/ekane

FUTURE TRAVEL VEHICLES BY EDWARD KANE

SPACE

- "Space 2020's: What's Up There?" - Kindle & paperback
- "Bargain Space Trips" - Kindle & paperback
- "Space Renaissance in the 21st Century" - Kindle & paperback
- "Search For Life in Space" - Kindle & paperback

TRANSPORTATION AND TRAVEL INNOVATIONS

- "Future of Transportation: 2020's and Beyond" - Kindle & paperback
- "Electric Vehicles for All" - Kindle & paperback
- "Hot Electric Vehicles for the 2020's" - Kindle & paperback
- "Important Innovations: Transportation" - Kindle, paperback & Audiobook on Audible
- "How to Travel in the Future" Vol. 1 & 2 - Kindle & paperback

LISTS OF TOP NEW INNOVATIONS

- "List of Best New Innovations" - Kindle & paperback
- "List of Top New Environmental Innovations" - Kindle & paperback
- "List of Top New Gadgets" - Kindle & paperback
- "List of Top New Medical Innovations" - Kindle & paperback
- "List of Top New Energy Innovations" - Kindle, paperback & Audiobook on Audible
- "List of Top New Robots" - Kindle, paperback & Audiobook on Audible
- "How to Use AI & AR" - Kindle & paperback

INVESTING IN INNOVATIONS

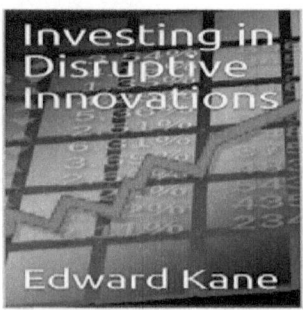

- "Investing in Disruptive Innovations" - Kindle & paperback

Fiction - Adventure, Life Lessons Book for Children

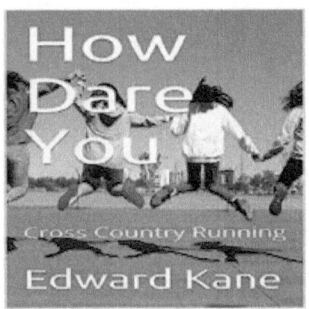

· "How Dare You" - Kindle, paperback & Audiobook on Audible

amazon.com/author/ekane